THE
PIT N' POT
GROWER'S
BOOK

The Pit n' Pot Grower's Book

Jack Kramer

Drawings by James Carew

THOMAS Y. CROWELL COMPANY
ESTABLISHED 1834

Designed by ABIGAIL MOSELEY

Manufactured in the United States of America

Library of Congress Cataloging in Publication Data

Kramer, Jack, 1927–
    The pit n' pot grower's book.

    Includes index.
    1. Container gardening. I. Title.
SB418.K76   635.9′65   74–23821
ISBN 0–690–00717–5
ISBN 0–690–00745–0 pbk.

1 2 3 4 5 6 7 8 9 10

Other Books by Jack Kramer

1000 BEAUTIFUL HOUSE PLANTS & HOW TO GROW THEM

GROWING ORCHIDS AT YOUR WINDOWS

GARDENS UNDER GLASS

PLANTS UNDER LIGHTS

FREE EARTH GUIDE TO GARDENING

EASY PLANTS FOR DIFFICULT PLACES

COMPLETE BOOK OF TERRARIUM GARDENING

PHILODENDRONS

FLOWERING HOUSE PLANTS MONTH BY MONTH

HOW TO USE HOUSE PLANTS INDOORS FOR BEAUTY & DECORATION

ORCHIDS FOR YOUR HOME

THE INDOOR GARDENER'S HOW-TO-BUILD-IT BOOK

For Barbara
and her enthusiasm

Acknowledgments

I PARTICULARLY want to express my thanks to Jack Barnich, who took most of the photographs for this book, and to James Carew, the artist who did the fine line drawings. Both of these men went far beyond the normal work limits to get exactly what I wanted on film and on paper for illustration.

I also want again to thank Judy Smith, who did typing on weekends to get this book out on time. And as always (as with every book), my thanks to the plants themselves for responding and growing for me, and for the joy they give me every day of the year.

Contents

8. Houseplants   87

9. Wild Plants   110

Introduction: Never Throw Anything Away

IF YOU KEEP throwing away avocado pits, pineapple tops, or pits and seeds from exotic fruits such as mangoes and papayas, you are throwing away free plants. If you don't snitch a few seeds from a friend's houseplants, or harvest your own, you are wasting money (note the high cost of houseplants today). And if when you walk through the woods you don't recognize seedpods waiting to be planted, you are missing more free plants from nature. And taking a few seeds will not harm the environment. Mother Nature is generous; work with her, not against her, and save fistfuls of money doing it.

Almost any seed or pit will yield a lovely plant for your home or patio, and though starting the plant takes a little time and growing it requires some attention, when you have nurtured it from seed to maturity it is indeed a joy. It gives satisfaction and solace for the soul as well as beauty for the eye.

Pit n' pot gardening may not give you mammoth plants to decorate every corner of your home, nor will they last a lifetime, but you will have some nice small plants to tend for years. And there is excitement in this activity, too. Have you ever seen a guava plant? Do you know what papaya leaves look like? Have you ever seen tiny begonias sprout, or watched the leaves of elderberry or bloodroot unfold? You are in for an adventure. Every day you will be checking your indoor garden to see just which plant is doing what. It is all part of the fun. Once you get started you'll be surprised at your green fingers.

Here is yet another facet of nature waiting to be explored. All it takes is a little know-how, and that's what this book is all about—helping you make unusual plants a part of your everyday experience (while saving money, too).

JACK KRAMER

THE
PIT N' POT
GROWER'S
BOOK

1.
What You
Should Know

A GROWING PLANT—even a small one on the windowsill—represents the miracle of life itself. To see it mature from a seed or a pit (we use the words interchangeably in this book) into beautiful green leaves is one of the joys of gardening. The pot on the windowsill will provide daily excitement and education for the family: Is the plant growing? How does it grow? What will it look like? You'll feel like a proud parent—when you're successful, that is.

There are many advantages in not-throwing-anything-away gardening. One is cost. Plants have increased in price, like all things, so it makes good sense to start your own. And it's easy; it's a simple matter to take the pits and seeds of various fruits (kitchen plants, as I call them) after they're eaten and plant them.

Another bonus of pit n' pot gardening is that plants become beautiful indoor accents. Everyone likes leafy plants for room decoration. For example, some apartments (often young people's, I'm happy to say) are veritable jungles, with the plants replacing more utilitarian things like furniture. (After all, one can always sit on the floor!)

Seeds or berries you collect from wild plants can contribute to your indoor garden, too. The plants may not last for years, but for a season or so they'll provide green accent, and if you have an outdoor garden, they can, when old enough, be planted there; otherwise, enjoy them for what they are.

Still another advantage of pit n' pot gardening is that you can grow unusual specimens you ordinarily would never see. A papaya or mango at the window of a city apartment is unique, as is a wild plant such as a yucca; you can't buy these plants at commercial florists.

So all in all, pit n' pot gardening makes good sense, whether for educational reasons, cost, decoration, or potential outdoor plants. For some energetic gardeners not throwing anything away may result eventually in a jungle in the apartment in a few years, but for most people it will provide a few choice plants for the home, beauty to the eye, and solace to the soul. What more could you ask of a tiny seed or pit?

To accomplish all this you don't need extensive or expensive equipment or tastes, merely curiosity and a little knowledge of plants. There's really very little mystery to it; the miracle is already in the seed or pit. What you have to know is how to get the seed started (germination) and then how to keep it growing. But remember that the more difficult the challenge, generally the more rewarding the victory.

The most popular pit n' pot plant is the avocado; pinched
at the base so that it will branch, it makes a splendid houseplant.
(PHOTO BY ROCHE)

The guava, a tropical fruit, makes a fine leafy plant for the home, too; it must be started from seed after you eat the sweet dessert. (PHOTO COURTESY USDA)

What You'll Need

For starting seeds and pits you will need containers or pots. You can use throwaways—tin cans and coffee cans—or the standard clay pot, which is really a fine container. You can use household items like shallow casserole dishes or four-inch plastic trays, sometimes called seed trays, available inexpensively at nurseries. Other throwaways that can be put to use as containers are mentioned in Chapter 5. No matter which container you use for seeds or pits, provide drainage holes so

Few people get to see a litchi nut bush; yet they eat the delicacy in Chinese restaurants. The plant itself is handsome, as shown in this photo. (PHOTO COURTESY USDA)

excess water can escape. Punch small holes in the bottom. Also put in some stones or broken pot pieces (potshards) to assist in drainage.

Peat pots (available commercially) are also used as seed starters. These are convenient because when seedlings are up and ready for planting, you plant peat pot and all and this lessens the transplanting shock. With other methods you have

Houseplants like episcias are excellent indoor plants and can be grown from seed collected from plants or purchased in packets. (PHOTO BY MATTHEW BARR)

At proper times, seeds of wild plants can be collected for starting
your own plants—not only a pleasant venture in growing but
nice to get out in the woods. (PHOTO BY MATTHEW BARR)

to pick out seedlings and transplant them, but I prefer it this way, and this may be simply because I have always done it this way.

To start seeds and pits in the containers you must provide them with a sterile growing or starter medium. These come in packages at nurseries and are discussed in Chapter 3.

When seedlings are up and ready for transplanting, you will have to put them in soil in new containers. There are as many different kinds of soil mixtures as there are recipes for fruit cakes, but the basic principles of a good soil are the same—it must be porous, full of nutrients, and not too clayey or too sandy. Clay soil can't absorb water, so the plant starves; sandy soil allows too much water to escape with little benefit to the plant. Thus, a soil in between the two is what you want.

Soil must have nitrogen to promote good leaf growth, phosphorus for good stem growth, and potash to help keep the plant healthy. (Trace elements are necessary, too; these are already in good soil, so you needn't worry about them.) Fresh soil has enough nutrients to keep a plant growing for at least six months. After that time, repotting in fresh soil is necessary. Don't mess with compost bins for just a few plants unless a friend has one you can borrow from. Pit n' pot gardening is for people who want to keep a finger in the soil but don't want to get bogged down in the mud.

For our purposes, soil testers, soil conditioners, and other soil amendments that are constantly being pushed by advertisements should be disregarded almost entirely. Buy, if you can, a good bulk soil mix (by the bushel) at a greenhouse or nursery. Or, for convenience, use packaged soils. Some are very clayey, so squeeze packages a bit (as you would bread) to be sure you are getting a spongy, porous soil because that is

Seeds of wild plants; these can be started indoors to get small leafy plants for the house. (PHOTO BY J. BARNICH)

the best. If you use plain old soil from your yard or someone else's yard, be wary. It could be good or it could be awful and harbor disease and fungi that can cause trouble for your plants. You can sterilize this soil by baking it in the oven at 220° F. with the door open, and so on and so forth, but it is an ungodly mess and smells terrible.

Questions

Many people ask about how long it will take an avocado pit to start growing, papaya seeds to sprout, and so on. Germination varies from plant to plant, and I have tried to mention germination times for specific plants in future chapters. But many times germination can be erratic and may occur later than it is supposed to. Don't give up too easily. What you think is lost forever may in six months' time start growing.

You may also wonder how to prepare seeds and pits from fruits. Some pits, especially mangoes, are large and fibrous, difficult to clean. Other seeds or pits (for example, those from papayas and pomegranates) are slick and slippery and must be freed of their protective coating before they are planted. Generally, clean the seed or pit of all flesh, wash it, and then thoroughly dry it. Air-drying for a few hours or overnight should be sufficient. Then plant the seeds or pits. Information on planting is given in Chapter 3, but here we show the procedures for some of the plants people are most curious about. Avocado pits are suspended in water. Coconuts are started whole, placed at an angle in a large container of growing

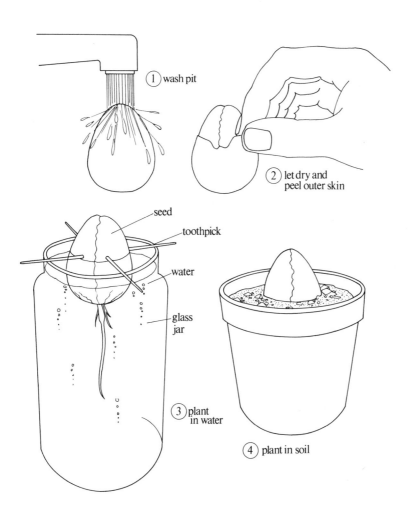

① wash pit

② let dry and peel outer skin

seed
toothpick
water
glass jar

③ plant in water

④ plant in soil

Avocado Pit

11

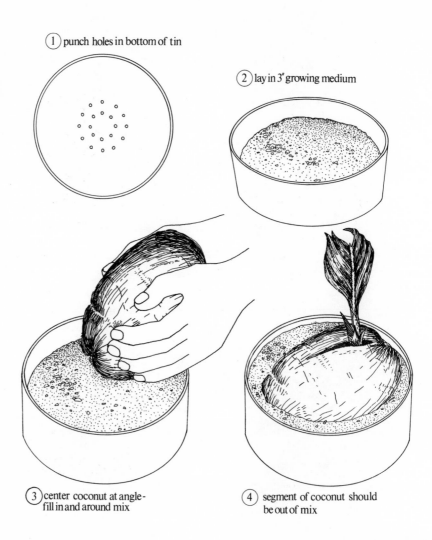

(1) punch holes in bottom of tin

(2) lay in 3″ growing medium

(3) center coconut at angle -
fill in and around mix

(4) segment of coconut should
be out of mix

Planting Coconut

① CUT HERE

② LET DRY A FEW DAYS

③ PLANT

sandy soil

clay pot

pot shards

WATER ④

Planting Pineapple Pit

13

medium. With pineapples you slice off the clump of leaves and include about 1 inch of the top of the pineapple. Let dry a few days and then put it in planter mix as shown in the drawing. All is explained in later chapters.

Many of the seeds from wild plants are much easier to prepare for the pot because they're in convenient capsules on the plants; all you have to do is crush the capsule, capture the seed, and plant it. Berries need special treatment. Remove the pulp and thoroughly wash the seeds in warm water and clean them before planting. Or throw the berries in a sieve and crush away the pulp; what's left is the seed.

Gathering your own houseplant seeds will furnish some preliminary sex education at home. You help nature along by touching one thing to another—the stamen, or pollen from it, to the pistil (see Chapter 2). Soon seeds form on the plant— that is, sometimes. Then collect the seeds and dry them and make them ready for the pot.

All the above is not really technically proper, but for our purposes it will work, and the pictures throughout the book will prevent you from getting confused.

2.
What You Don't Have to Know (But It Will Help)

YOU NOW KNOW some of the things you should know, so it's time to know some of the things you don't have to know but that will help. Indeed, if you go on from these humble beginnings of windowsill gardening to outdoor gardening, this chapter may be the one chapter you'll read again. Here is where we explore what seeds are, species and genera, and other little-known facts.

How It Starts

The ultimate purpose of flowers is seed production, to perpetuate the species. Growing a plant from seed is known as

15

the sexual method. You can also grow plants by asexual, or vegetative, methods (cuttings, layering), but with a couple of simple exceptions, this is beyond our scope here.

Two organs in flowers produce a seed—the stamen, which has pollen grains that form the male cells, and the pistil (female organ), which is generally in the center of the flower. The five steps leading to the formation of seeds or pits are:

1. Opening of the flower.
2. Transfer of the pollen from stamen to pistil.
3. Germination of the pollen.
4. Fertilization.
5. Growth of the fertilized egg into an embryo with a surrounding case (the seed).

Seeds are produced differently in different plants. One group (gymnosperms) produces seeds without any protective coats; in the other group the seeds develop within a vessel (ovary). The ovary is that part of a flower that later becomes a fruit with seeds inside it. The ovule contains a sac and tiny eggs. An egg can be fertilized by a sperm cell (pollen) that travels in a pollen tube that runs down into the ovule. After fertilization, the resulting embryo contains a special storehouse of food for its own use after it is separated from the mother plant.

Gymnosperms have some protection, too, during development. Seeds of pine trees, for example, are protected by the scales at the base of the cones; the scales separate at the proper time to release the seed to the air.

Plants from seeds of cultivated fruit trees and ornamental garden flowers don't come true to variety. That is, they can't reproduce a duplicate plant, so these plants are grown vegeta-

This citrus seedling started from seed is being transplanted in
its original peat pot into a large clay pot. Soil will be added to
the top of the pot. In a year or so, the plant will be a handsome
home addition for living room or any room. (PHOTO BY J. BARNICH)

tively from cuttings, layering, and so forth. Because the sex
cells of these plants carry random assortments of mixed-up
characters, the resulting plants are unlike those of the varieties
that produced the seed. (Citrus, however, is an exception because no cross fertilization is involved; the seeds from citrus
therefore come true to variety.) But plants from seeds of species (wild plants) breed true if precautions are taken in pol-
linization.

The miracle of seed growth is shown in this special photo. If you look closely you can see seed, root, and the first flush of leaf growth. (PHOTO BY J. BARNICH)

Where Plants Come from

Plants come from virtually all parts of the world, and many of the plants you may wish to grow may have come originally from a very different climate. The avocado and papaya are native to Central and South America and have been used for centuries as food plants. Pineapples were introduced to Europe in 1555 from Brazil and Paraguay. African violets, of course, were originally from Africa, and begonias from South America. Among our native plants are azaleas, rhododendrons, and hydrangeas. Persimmons are native to China, and mangoes come from India. Yet because plants are adaptable, they will adjust to other environments, within limits. Persimmons do grow in America, as do mangoes and azaleas, and rhododendrons are grown in Europe. If you can duplicate the native conditions of a plant, it has a better chance of growing well, so knowing the country of origin of a plant helps you understand just what it needs in the way of temperature and moisture. (In our plant sections we mention some of these places of origin.)

Here life starts for the wild plant lunaria as leaves unfurl. (PHOTO BY J. BARNICH)

Most garden plants are hybrids; that is, they're the best mated with the best to produce outstanding traits—flower size, flower form, or resistance to disease. Some houseplants, too, have been bred extensively, resulting in outstanding varieties.

In order to germinate, most seeds need good humidity. This can be supplied by placing a plastic cup or glass jar over the seed in its container, or cover the entire container with plastic. (PHOTO BY J. BARNICH)

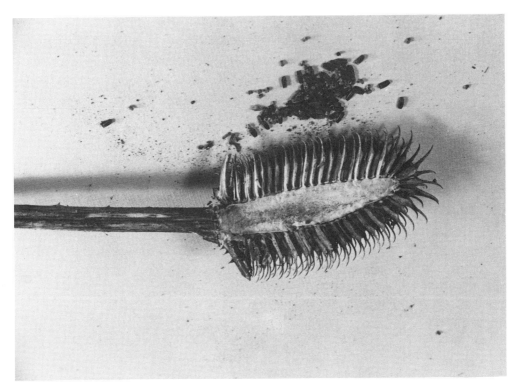

Seeds as they come from the pod. Sometimes you can shake them loose; other times they must be separated by rubbing the seed container or threshing it. (PHOTO BY J. BARNICH)

Plant Names

Most people know an avocado when they see it, but if they were asked its botanical name, they'd have to shrug their shoulders. There's really very little reason for knowing the botanical name, and yet it offers some interest. The avocado pear is correctly known as *Persea americana. Persea* is the group name and *americana* its species or individual name. Since there are about fifty different species of *Persea,* you can see why botanical names are important.

For the windowsill gardener, knowing the common names of plants is generally sufficient, but if this book spurs you on to bigger and better gardening, it's wise to know a little bit more about plant names. The problem with common names is that in one area a plant may be known by one name but in another part of the country by something entirely different.

The botanical tongue twisters that most people avoid are difficult to pronounce but not difficult to learn if you really have the desire to know plants. And calling plants by their botanical names is sometimes the only way always to get the seed you want (if you buy by mail order, for example). With plant names, the group name (the genus) is similar to your last name, and the species name is like your first name. When hybridists create an improved variety, the name would be somewhat like a "Jr." name to us.

How Plants Grow

Seeds start to grow because of the right life-starting combination of water, air, temperature, and age and stage of maturity. Too much heat may cook seeds, and cold and frost may injure some seeds. If there is too little moisture, the seed's plant foods will not be in the necessary solution; however, too much water will rot the seed. An evenly (and moderately) moist soil will start the germination process. Germination requires the

Seed pod in the wild. Crushing the capsule or shaking it will release the seed. Then it can be started in shallow containers of vermiculite or other "starter" mixes. (PHOTO BY J. BARNICH)

seed's absorption of moisture, the proper temperatures to transform stored food into sugars by enzymes or natural ferments, and the bursting of the seed coat.

Your Own Seed— Pollination

Most gardeners buy houseplant or wild seed already processed and packaged from suppliers. But of course you can take seeds from your own plants. In this case you must know a little something about sexing (breeding). The operation in plants, as it is in humans, is quite simple.

The best time for pollination is when the flower appears mature (although even when the flower falls off, usually both pollen and stigma are still ripe for fertilization). At maturity, the tip of the pistil, the stigma, becomes slightly sticky, enabling it to hold the pollen. To cross two flowers you must transfer the pollen from the stamen of one to the stigma of another; there are several ways you can do this—it's only a matter of using whichever method seems easiest. If you cut a tiny section in the pollen sac on the stamen (the sac is called the anther), and let the pollen fall on your thumbnail, you can then place this pollen on a stigma of your chosen seed parent. Another simple way is to take an anther from a flower (even one that has just fallen off), slip open the pollen sac with a needle, and then gently pinch the sac to open it wider and apply the opening to the sticky stigma of the seed parent. A small artist's paintbrush can also be used to transfer pollen to a stigma. The best time for making your crosses is during the middle of the day when the air is warm.

Collecting seeds in the wild offers adventure—the lure of the unknown. Generally, when you collect your own seeds, do it when they're ripe, and then you can clean and store them in a cool place until you sow them. Some types, however, have to be gathered before they're fully ripe to prevent dispersal. For example, fully matured pine seed cones can open rapidly on a warm spring day, in which case seed is lost.

3.
What You Had Better Know

Growing your own plants from seeds and pits is fun. Who doesn't like something for nothing? But although there's little cost involved, some care must be given to the seeds and plants.

Getting Started

When your first attempts at pit n' pot gardening start to yield freshly sprouted leaves, you're likely to bound off to the nursery for plant foods. Don't do it; those tiny specks of green don't need food now, and you haven't yet become a gardener. Photosynthesis doesn't start until true leaves appear, so don't use fertilizer until it's actually needed. When the first true leaves appear, and you'll see them, food then becomes necessary. But until that moment, furnish a constant supply of moisture and proper temperatures. In other words, be careful and precise about getting seeds started, and have patience.

Seed may come from your own plants (I hope you try this method), or if you want a shortcut, get them from supply houses. To get your own seeds from plants, remove them from cones, pods, or fruits. Prepare them for immediate sowing, or you can store them for future use. When storing seeds, be sure they're dry and clean and free of any pulp; how you store seeds makes a difference in how long they are viable, that is, endowed with life. Many seeds remain good for years if properly stored. The best method is to put the seeds in an airtight, covered jar, label them (or you'll never remember), and place the jar in a cool, dark place—a garage or basement is fine. Keep the seeds dry and be sure they're not subjected to temperature extremes.

Germination Problems and Cures

Nature is generous; she has little tricks, but you can work with her. Some seeds are difficult to germinate because their coatings are thick, but it is easy to soften the outer coating to speed germination along by soaking the pits or seeds overnight in warm water. Then the tiny plant can break its shell without

1. put fruit in strainer

2. run water thru strainer on fruit

3. remove pulp- leaving seed in strainer

4. dry seeds in foil at 200°F

5. store in sterile glass with lid

Separating Seeds From Fruit

Seeds will be started in this special starter mix in a plastic throw-away container. (PHOTO BY J. BARNICH)

undue labor. Put the seeds in hot water (180° to 210° F.), and allow them to soak for twenty-four hours as the water cools. Don't leave them longer; they'll be soggy. And don't boil them. Sow the seeds immediately after the soaking process; if they dry, the treatment is likely to be a failure.

Other plant seeds need a cooling period before they are sown; this is known as stratification. Don't let the word scare you. It simply means the seeds need coolness for a period of time (generally 60 to 100 days). Seed can be mixed with damp sand, shredded sphagnum, or a mixture of peat moss and sand. (Be sure the medium is moist but never saturated.) Wrap very small seed in cheesecloth first. For a container use a jar (with a lid), a seed tray (with a plastic cover), or simply a Baggie and then tie it. Now place the container in the refrigerator and label it and be sure an unwary guest doesn't raid your icebox.

To assure good humidity for germination of the seed, a glass jar is set on top of the container. (PHOTO BY J. BARNICH)

A good way to provide adequate humidity for seed sprouting is to put plastic over the container. (PHOTO BY J. BARNICH)

Another trick to get seeds started is to nick the edge with a sharp knife. This breaks the hard coat and allows the seed to germinate more readily than if no cut were made. This process is called scarification.

Water, temperature, air, and light are necessary for germination. Bright but not sunny light is best, and warmth (70° to 80° F.) is fine.

In Chapters 7 and 9, where care of individual plants is discussed, you will find more information on helping specific seeds germinate.

Sowing and Starter Mixes

To prepare the containers for the seeds or pits, fill them to within one-half inch of the top with a sterile mix. Moisten and press the material down to eliminate air spaces. Cover large and medium pits or seeds with a dry layer of mix (twice the thickness of the seed is the general procedure). Scatter fine seed such as begonia seed on the top of the growing or starter mix and then sprinkle with some more of the medium. Now apply a thin sprinkling of water and provide some sort of cover to hold in moisture such as a tent made by placing a plastic Baggie on sticks stuck in the pot to cover the container (see Chapter 5). Check daily to be sure the growing medium is adequately moist, but never wet.

Seeds need a sterile medium in which to germinate, and it is best to use one of the commercial packaged "starter mixes" or growing mediums available at nurseries. Because there are so many I will list some here:

VERMICULITE. This is an expanded mica that holds moisture well. It is sometimes available under the name Terralite.

These seedlings are just starting life; they will need good warmth and moisture to survive. (PHOTO BY J. BARNICH)

MILLED SPHAGNUM. This has been used for many years and is finely milled sphagnum. It works very well and is highly recommended.

PERLITE. Perlite is a volcanic ash that has been used more frequently lately than before. It tends to absorb moisture rather than hold it and stays quite cool—which can be good or bad, depending on the type of seed sown.

VERMICULITE-SPHAGNUM-PERLITE MIX. This is an excellent mixture for starting seeds. It's light, clean, easy to handle and holds moisture but readily drains. To my knowledge this mix is not available already mixed in packages. You will have to do it yourself, which is not a difficult task.

SOIL. Soil, packaged or otherwise, can be used for seed, too, as long as it is sterile, and packaged soils are. You might want to add some sand or peat to the soil because it is imperative that a growing medium for seeds drain readily and yet hold some moisture.

Large seeds, such as citrus seeds for example, need to be planted somewhat deep; many large seeds need to be nicked at the edges (scarification) to germinate or be soaked in water overnight to soften the seed coat. (PHOTO BY J. BARNICH)

seeds in growing medium

shallow pot

1″ clay saucer

mist seeds

Ways of Watering Seeds or Pits

Care of Young Plants

Once sprouted, seedlings will die if they become dry, so provide even, moderate moisture at all times, that is, daily if necessary. Mist with a hand sprayer or put water in a saucer under the pot and let the mix soak up the moisture. Not enough water causes plants to die, but too much water reduces aeration in the medium and leaves plants prone to damping-off, which neither you nor the plant wants anything to do with! Do not water plants with icy or hot water; room temperature water is best. (Let a bucket of water stand overnight.)

When the second set of leaves emerges, move containers to more light (perhaps some sun), and if you have used a plastic or glass covering, remove it. Keep the medium uniformly moist. Weak fertilizer such as Hyponex or Rapidgrow can be used now once every two weeks.

When seedlings are about one to two inches tall and have separate leaves, transplant them to pots with regular soil. To remove the seedlings, lift them carefully with a blunt-nosed stick or your finger, getting as much of the root ball as possible. Water pots of soil well with a fine mist spray. Give them some sunlight now. Be sure the soil doesn't dry out.

4.

Where to Get
Seeds and Pits

THERE ARE NUMEROUS PLACES to get the thousands of available plants. The easiest method is to buy a mature plant, but why do it unless you have to? They're expensive, and I've noticed that my friends have to buy new plants every few months to replace the ailing ones. People who plant pits from fruits they eat or gather seed from the wild derive much more enjoyment, and window plants eventually become living-room (decorative) plants as they grow to maturity.

Kitchen Plants

The first part of the plant section of this book is called kitchen plants mainly because the kitchen is associated with food and

Your mango tree will not grow this large nor bear fruit, but it will still be a pleasant small plant for indoors. (PHOTO BY USDA)

The lovely leaves and bushy growth of the persimmon are
evident in this photo. Small plants can decorate the indoors, too.
(PHOTO BY USDA)

most of these plants are food plants or fruits. After eating the fruit why not plant the pit or the seed? Of course, as with all plants, unusual fruit plants need a few special considerations before they'll grow and put forth lovely leaves. For example, litchi nuts, those delicious Chinese delicacies, can't just be planted because in most cases you'll be eating a dried litchi nut, not a fresh one, and only fresh ones bear viable seeds that

Begonias are favorite houseplants and can be grown from seed—less expensive and more satisfying than buying mature plants. (PHOTO BY MATTHEW BARR)

have a chance to grow. However, any good Chinese store will have fresh litchi nuts in summer. With papayas, Mother Nature's little trick is that fresh papaya seeds, when planted, soon become spoiled seeds, which is not the papaya's fault. It's yours. The seeds must first be extracted from their coatings and then be washed and dried—no easy trick. The best method seems to be by sieving pulp from seed. Pomegranate seeds must also be separated from the fleshy edible pulp, and with mangoes the large pit must be scrubbed and scrubbed, again and again, before it is ready for planting.

The most unusual fruits aren't available at all times of the year or in every store, as I discovered when I wanted to

Pomegranate seeds being separated from the sweet-tasting pulp. (PHOTO BY J. BARNICH)

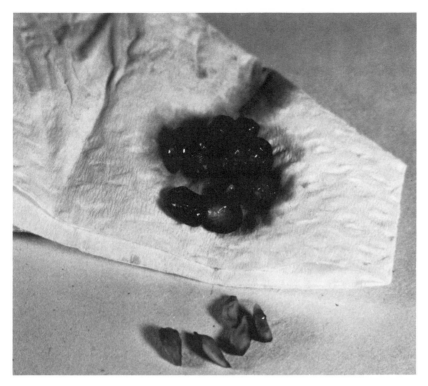

Pomegranate seeds after the fleshy pulp has been removed. Now they can be planted. (PHOTO BY J. BARNICH)

start some seeds in the course of doing this book. But they're available in season in most large produce departments or grocery stores. Buy them and eat them in season for full enjoyment. Don't throw the pits or seeds away if you cannot plant them at that moment. Wash and dry the seeds and store them in dry glass jars in a cool place.

The more unusual the fruit or houseplant the more fascination it holds, for me at least, because it gives me a chance to see something I've never seen before and provides me with an adventure in growing plants. For example, I never expected the beautiful litchi plant I got.

If you make a few mistakes the first time and nothing but weeds is your reward after making a planting, don't give up. You're wasting little by trying again, and sometimes coincidence will be the key to success and plants will grow. Years ago in Florida I attempted to grow a coconut palm. I tried over and over with little success until one day I didn't have a suitable container. I used what was at hand—an empty shallow tin can. The coconut was too large for the can, but I was determined, so I planted it, with half the coconut above the soil line and protruding at an odd angle. This coconut sprouted. I've since grown several coconuts this way. An avocado I started in Chicago and took with me to California now grows in my yard. It's seven feet high, and although I've never had the fortune to have it bear fruit, it's still a lovely plant for the garden. Patience is a virtue that really applies to gardening.

Thus, for those of you who have greenhouses or backyards, the pit n' pot method is just the beginning of bigger and better things. Even without a backyard or greenhouse, pit n' pot gardening still affords lovely green plants for your window or living room; it's not everyone who has a papaya growing in his living room!

Houseplants

Most houseplants are purchased, but if you want the joy of growing things, starting your own plants from seed offers more enjoyment. As described in Chapter 2, just pollinate flowers by hand and harvest seed when it's mature.

The delicious fruit of the mango is well known; the seed or pit is lesser known. It must be washed and scrubbed (a devil of a job) before it can be started in the growing medium. (PHOTO BY J. BARNICH)

You can also buy seed already packaged, but in some cases growing your own seed is much more sensible, especially if you have a favorite plant you want to be sure of never losing. Once it bears flowers, do the pollination; you'll have enough seed for future generations. Some houseplants, including begonias and gesneriads, develop seed readily by hand pollination. So don't depend on the local florist to have every favorite houseplant—in most cases he won't, anyway—do your own. In the process you'll gain knowledge and a certain satisfaction, too.

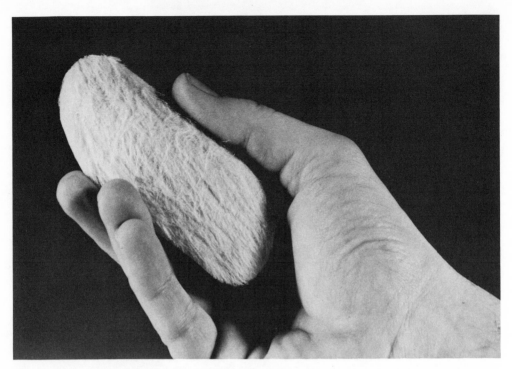

A dried mango pit ready for potting. (PHOTO BY J. BARNICH)

Wild Plants

Conservation-minded pit n' pot gardeners will appreciate the wisdom of growing wild plants from seed. Many of our beautiful native plants are now protected by law and can't and shouldn't be picked. But taking some seed from them isn't against any law I know of, and I think it helps nature. Plus, you'll have better plants from seed than by stripping plants from the countryside. You'll also experience a new world because plants such as Dutchman's-breeches, hepatica, spring beauty, and little lady's slipper orchids are intriguing, and the seeds of wild plants come in many parcels and pods, in weird and exquisite shapes that would otherwise escape your eye.

Transplanting wild flowers is a difficult chore, but growing them from seed—although care is needed—isn't. The main thing is to gather seed at the right time of the year when they're mature and ready for planting. (More information on growing wild plants from seed is in Chapter 9.)

You can also buy wild flower seed from suppliers, but walking the woods or forest, visiting nature, and gathering your own seed for sowing is to me a pleasure. For those of you who can't make it to the woods, by all means try wild flowers from commercial seed sources because how else are you ever going to see these lovely plants? Either way, there's a treat in store for you because wild plants are nature's bounty, untouched by man and never hybridized, and can grow for you in a pot on the windowsill for many months.

5.

What You Can Do to Help Nature

SOME SEEDS, for example, grapefruit or lemon seeds, can be put directly into shallow pots of soil. They need neither special preparation nor special containers and usually will reward your push-and-plant method with—in time—leafy green shoots and satisfaction for all concerned. Other plants don't yield their magnificence in such an easy manner. Nature has been cautious, providing the seeds with protective coatings that, depending on the plant, must be dried and cleaned in various ways. And even though many seeds or pits don't need exacting cultural requirements (temperature, humidity), some do, so it's good to know just how to help nature along.

Some seed may require a propagating box (seed tray). Don't let the word scare you; think of an incubator, for in essence that is what a propagating box is—a container with a

46

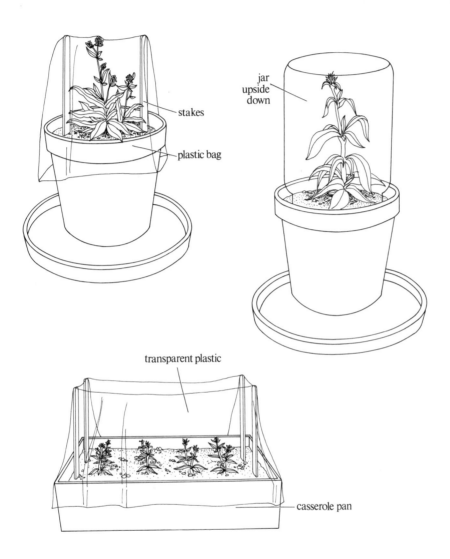

stakes

plastic bag

jar
upside
down

transparent plastic

casserole pan

Various Ways of Providing
Humidity for Seeds and Pits

lid or top in which you can provide controlled temperature and humidity to get your exotic babies to flower. Basically, a propagating box is a tiny greenhouse. The advantage of propagating boxes over pots is that in a larger container you can start more seeds, so your chances of success will be greater; also, air, temperature, and humidity and moisture are more easily controlled in a homemade propagating case than at a windowsill.

What to Use

For your propagating box use throwaways such as the aluminum pans that frozen rolls come in, discarded glass casseroles, or even cottage cheese or milk cartons cut lengthwise. Remember to provide drainage holes, so excess moisture can escape. And the design needn't be elaborate because once seeds germinate and sprout, growing true leaves, you can transplant the seedlings to pots of soil for your windowsill growing. From such humble structures you'll soon extract glorious plants.

To ensure humidity and warmth cover the tops of the containers with plastic or glass (a glass jar or plastic bathroom glass may do) or a Baggie propped on sticks. Commercially made, small-table greenhouses of plastic are also available, with heating cables in the bottom to provide precise temperatures. These are perfectly fine to use, but you can make one easily with throwaway materials.

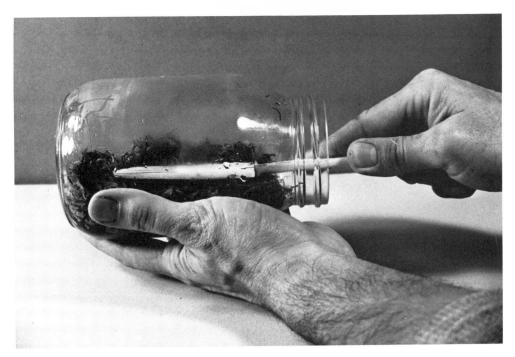

A fine way to start seeds or pits is to use a mayonnaise jar filled with sphagnum or starter mix. The jar provides excellent humidity for germinating seeds. (PHOTO BY J. BARNICH)

Where to Put Containers

Once you have your seeds or pits in their containers, and if you think bottom heat is in order, place the containers on top of the refrigerator to provide some bottom warmth. (But do not put containers on a radiator—too much heat may harm plants.)

Another good place to start plants may be the bathroom. Some bathrooms have translucent glass that gives you privacy and yet provides the right kind of diffused light for seedlings. Also, if you shower regularly, your plants will have ample humidity; even if you don't want to be scrupulously clean, run the shower for a few minutes a day to provide that extra humidity for the young plants.

Thinning the Jungle

Invariably, from little seeds and pits great trees grow, and there may come a time when you'll be so successful (with the aid of this book) that your plants will become too tall for the kitchen and require a place of their own. Depending upon the kind of plants they are—house, fruit, or wild—they can be moved into the living room for decoration or placed outdoors in the ground to grow to maturity, or, in the case of fruit plants, to fruiting, provided you have some cooperative bees.

Let's take a fruit plant like a loquat or persimmon and follow it through its life. Once it's too large for the original pot you started it in, transplant the plant to the garden area. Try to do this in early spring so the plants have a good chance to make growth before very hot weather. To remove the plant from its container, tap it against the edge of a table. Loosen the plant by grasping it by the crown and juggling it gently. Now dig a deep hole in the garden for the plant, insert a mound of soil, and station the plant. If it's too high in the ground, take away some soil; if it's too low, add some soil. Fill

Here a standard clay pot is used for pit n' pot starting; seeds are carefully imbedded in the starter mix. (PHOTO BY J. BARNICH)

in and around with soil and water thoroughly. For the first few weeks, keep a close watch on the fledgling to see how it's doing.

Artificial Light

Because nature is so generous with her wares it's only fitting that occasionally you help her to achieve her ends and yours—

Here African violets grow under artificial light; at the left end of the light tray is a glass bubble bowl for starting seeds under light. (PHOTO BY MATTHEW BARR)

Seedlings may also be grown under artificial light in a tray setup as shown here.

strong, healthy seedlings. You can do this by using artificial light to help start them germinating. This may seem a little like cheating, but for folks in cities like Chicago and New York, where winters can be mighty gray, this chapter may come in handy. When you're praying for the papaya pits on your windowsill to burst into plants on a January day, artificial light will be of great help.

There are elaborate artificial light setups—trays and carts —but for our kind of frugal gardening let's leave progress far behind and just concentrate on the plain old light bulb you read by. A 75-watt bulb directed on the plants or seedlings will furnish them with some light during gray days and hasten growth. But do not keep plants too close to the lamp because too much heat can quickly desiccate our little gems. A safe distance is thirty inches. Leave the lamp on from twelve to fourteen hours daily.

If you really want to delve into artificial-light gardening you can use fluorescent tubes; they aren't too expensive and won't throw your budget astray too much. Used reflectors (industrial kinds) can be purchased at auctions instead of buying new ones. Fluorescent light is somewhat better than incandescent because the light is evenly distributed rather than being concentrated, as with incandescent light. When using fluorescent tubes, keep seedlings four to six inches from the light source and use ten watts per square foot of area.

Fluorescent tubes come in warm white, cool white, and so forth; for our purposes warm white tubes are fine. Lamps specifically designed for plant growth, such as Gro Lux, are also quite suitable and supposedly superior to standard fluorescent light, but I've yet to notice any difference after using both kinds for many years, and regular tubes are less expensive than the special ones.

6.

What You Would Rather Not Know

So far, I've made it seem that growing plants from seeds and pits is all joy and little sadness. Well, all things beautiful must have some gloom, and nature isn't about to let you have your own little jungles without some fuss. The fuss amounts to the unwelcome crawlies and creepies known as insects and occasionally a downright social disease known as fungus. However, there are cures, so all is not in vain.

Uninvited Guests

Aphids, probably your most unwelcome plant guests, appear in many guises. Some are green, others are red, or black, and so forth. Aphids are soft-bodied, one-eighth-inch oval bugs that can multiply faster than rabbits. In a few weeks you can have thousands, so once you spot them, eliminate them fast. Don't go out and buy poisonous sprays that cost money, wreak havoc with soils, and smell up the house. There's an easier, old-fashioned way to get rid of them. Take some laundry soap (bar soap, not detergent) and dilute a hunk of it with water until you have a frothy broth. Put the solution in a spray bottle and go to work; shower the pests every other day for about a week, and eventually they'll either give in or move on to brighter and better fields—outdoors, for instance.

Another little troublemaker in your own garden will or might be mealybugs, which are white, cottony insects that form snowy mounds in leaf axils or accumulate along leaf ribs. They're ugly to look at, but they're not invincible. Keep mealybugs in check by dousing them with rubbing alcohol on cotton swabs. Do this every other day to really get them saturated and disgusted enough so that they'll give up and die.

Red spider mite is the bane of many indoor garden plants. This almost invisible insect can raise havoc. Because they're so minute they're difficult to combat, but they do leave clues. If you inspect plants afflicted with spider mites, you'll invariably see their tiny webs at stem axils and underneath leaves. The first line of defense is a strong water spray from the tap. If they persist and you have no patience for daily water misting, you might have to resort to chemical warfare, which means an insecticide called Dimite. Use it only as directed and with caution and keep it away from children and pets.

Another unwanted plant guest is scale. They look like little lumps of gray, white, or brown and attach themselves to plant parts. Scale are tenacious devils and can harm plants

Mealybugs occasionally attack indoor plants and can easily be seen. Get rid of them as soon as you see them and before they gain a foothold. (PHOTO BY MATTHEW BARR)

considerably, so be equally tenacious in combating them. The cures take patience: Pick them off with a toothpick, one by one; try a soapy solution to spray them away; or if the attack is vigorous, make a solution of warm water, soap, and remnants of used cigarette tobacco. Let it steep a few days and then soak the scale pests. Tobacco is bad for people and equally obnoxious for scale. They'll leave soon enough after a few inhalations of the tobacco solution.

The above don't constitute all the pests that may attack plants, but generally there's little sense in worrying about the offbeat intruders such as thrips and whiteflies or other unidentified insects that may occasionally show up. If such infestations occur, consult your local agricultural extension service; they'll help you to identify bugs and suggest cures.

Scale, small armored-tank insects, can be seen on the stem of this houseplant (*Chlorophytum*). (PHOTO BY MATTHEW BARR)

Above all, remember that generally insecticides have no place in the home, not only because of their poisonous residue, especially when used in the kitchen where food is present, but also because of the plants themselves. Young plants are more prone to chemical exhaustion from poisons than mature ones, so it's a good idea not to spray or douse unless you deem it absolutely necessary. It's actually best to start new plants.

Your Fault?

Many times, when a plant fails, the problem may simply be poor care on your part; you've done something wrong, or haven't done something you should have, or the plant has been subjected to unhealthy conditions. So if you see brown leaves and stunted growth, look to culture first.

For example, if the foliage turns brown or yellow, it could be because the plant is in too much heat and doesn't have enough air circulation. Leaves can drop off for no apparent reason because of watering with icy water or low humidity. When leaves turn pale in color and the plant shows weak growth, increase light and provide more humidity. And if humidity is too low and heat too high, some plants may show

Most insects accumulate at leaf axils and along veining as shown in this photo. The culprit is mealybug. (PHOTO BY MATTHEW BARR)

crumbling foliage. Leaves that turn gray or watery may have bacterial blight; rather than fool with this illness, discard the plant and start again. Finally, if foliage becomes coated with white, a sign of mildew, dust the leaves with sulfur or charcoal.

Damping-off Disease

If this tirade on insects has left you breathless and you feel you've had enough, bear with me for a little longer. Insects are really a minor problem and can easily be eradicated, but a soil-borne disease known as damping-off will always be hanging around your newly potted plants. Using a sterile mix goes a long way toward preventing damping-off, which is a disease that starts from too much moisture and not enough light. The tiny bacteria finally develop and take hold. When you see the seedlings seemingly rotting away at the soil line and collapsing, you have experienced your first sight of damping-off.

Rather than discard the plant (as suggested for some diseases), use a cure. Semesan is the best known of the disinfectants on the market, although Captan is also satisfactory. These are chemicals, but they're easy to use—just a pinch of powder —and won't contaminate anything in the kitchen. Once again, however, keep products out of children's and pets' reach.

If you'd rather not use chemicals in the home, try another alternative: You can give the plants more air and less humidity. But never, never use overhead sprays on seedlings because the small plants might not dry out by nightfall and lingering moisture and high humidity are the invitation to damping-off.

One last word about caring for plants, insects, and diseases. Often a plant simply doesn't take to certain conditions, but amazingly enough, if you move it to another location it may perk up considerably. The new conditions may be warmer or cooler or just what the plant wants. In its previous place the plant may have been in a draft, too close to a door, and so forth, so the policy of moving plants around a bit has merit. Try it if you have a seedling that simply sits and sulks.

Fungus has attacked this plant with mushy and brown markings at leaf edges and tips. (PHOTO BY MATTHEW BARR)

7.
Kitchen Plants

FOR THOSE SEEKING the unusual plant to decorate a window, there's no better way to get one than by potting the pits of fruits you eat at home. Start with grapefruit and lemon pits; eventually you'll want something more exotic. Seeing plants you've never seen before, such as papaya or mango, is thrilling, and even if you're not successful, what have you lost? The answer is "Nothing," and you just might get a lovely plant. Indeed, you probably will because the odds are all in your favor. If you plant many seeds—and I always do—the chances of something sprouting are pretty good.

Of course, half the fun of planting kitchen plants is eating unusual fruits. There's infinite goodness in papayas and mangoes, guavas, and fresh litchis. In fact, if you never start this journey with pit and pot, you might never have the opportunity to taste these delectable goodies. So either way, growing what you normally throw away has its merits.

Some of the following plants can be started in a pot of soil; others will need more elaborate but not expensive devices such as propagating cases (explained in Chapter 5).

The kitchen plants you're successful with won't be with you for years. True, a few may become permanent houseplants —pineapple, lemon, grapefruit—but the majority, such as mango or papaya or tropical plants with exacting requirements, will eventually need outdoor culture, so don't be misled or misdirected. These kitchen plants, which cost you nothing, will be around for a few months, perhaps a year, after they start growth. If you live in an all-year temperate climate, many of the kitchen plants, including the avocado, the mango, and the papaya, can be transplanted outside and grow into lovely trees with no more work on your account. Many a pit n' pot gardener has been pleasantly surprised when, in time, his windowsill fledgling became a stalwart tree.

Avocado

The lush green meat of the avocado has furnished many a table with salads, and its rather heavy rounded pit has furnished many households with a lush green plant. Part of the avocado's popularity is due to the fact that the pit is easy to get started. There are several methods of approaching the birth of your avocado. Some people (I'm one) merely clean the pit and put it in a glass jar half full of water. You can also prop the pit on toothpicks as shown on page 11. In a few weeks it's off and growing roots; I then plant it in soil.

To be more certain of germination, however, cut a thin section from the apex and the base and peel away the papery pit coating. Put the pit in the soil or water with its base downward—that is, the broadest part of the pit. If in soil, don't embed the pit too deeply; cover it with about one-half inch.

Be sure the avocado has sufficient drainage; although it isn't choosy about soil it is particular about stagnant water at the roots. The plant has a tendency to shoot straight upward, so once it is growing well, clip off the top to encourage side branching or it will get leggy and unattractive. Even so, you may have to stake the plant to keep it attractive.

Repot the avocado frequently (every six months); each time put it in a larger pot. Eventually you'll have a handsome tree that will outgrow the windowsill and be suitable for a living-room spot. Avocados are formally known as *Persea americana*.

Chinese Gooseberry

Don't let the name throw you; what you'll be looking for is the kiwi fruit. The true gooseberry (*Actinidia sinensis*) is native to China, but the fruit we find at local markets is from New Zealand. This lovely twiner, with its fuzzy leaves, is ideal for the windowsill. The fruit itself is the size of a small egg but is shaped like a gooseberry and is chartreuse in color. When sliced and sprinkled with lemon the fruit is somewhat puckery but good; it's not an all-time favorite but worth a try. Don't eat the seeds in the center; plant them. Pick out the small seeds from the center of the fruit and dry them on a blotter or newspaper.

Because the kiwi is subject to our old nemesis, damping-off, sow the seed in a coffee can or flat azalea pot and cover it slightly with vermiculite to protect the stem-to-ground part. It takes about eight weeks to germinate. If you have trouble sprouting the kiwi pits the first time around, give them a cooling-off period in the refrigerator before the second attempt. To do this, mix the dried seed with some sphagnum moss in a Baggie and store at 45° F. for about forty days. (This process supposedly stimulates the natural cycle the kiwi goes through in nature.) Then replant the seed in containers (coffee cans or shallow pots are fine). In the latter process you'll have sprouts within three weeks; transplant them when they're a few inches high. If you use the first method, pray hard.

Citron

I found this delightful fruit in Chinatown during the winter season. Citron, botanically known as *Citrus medica,* is, under ideal growing conditions, a dwarf tree of about eight feet with a large (six- to eight-inch-long) ovoid fruit. The fruit, which is rough-textured and fragrant, is the citron peel used for fruit cakes. There is really no need to grow this one for its fruit, but the plant itself is rather pretty, with leafy green toothed leaves. Meticulously peel the citron and reserve the peel for

Leaves

Fruit

COCONUT

future use. Wash away all pulp and plant the seed or pit in some sandy soil in a shallow container. Bury the seed about one inch deep in the soil and keep warm (72° F.). The sprouts should start in about a month. When large enough, about four inches tall, put the seedlings into individual pots with soil that contains some calcium. The citron is an evergreen tree, so be sure to keep watering it through winter, although in these months temperatures can be somewhat cool (50° F.).

Coconut

If you have never tasted the white meat inside a fresh coconut, you are missing something. It is good. And if you have never grown a coconut palm in your kitchen, you are missing something equally good. It is a handsome plant that lends a tropical note to gray days.

The nut itself has a hard outer shell or hull. One end is rounded, the other narrow. It is from the round end that the young shoot emerges. With coconuts you need two, one to eat and one to plant, because the whole nut, *hull and all,* must be used, and this is rarely the way they appear in supermarkets, but you can find them sold as novelties at various airport boutiques, especially in Florida where I got mine.

You will need a large tub to start your coconut, and I used a half barrel with a rather poor soil and sand base mix as a starter. Do not bury the coconut; all you will have is a buried coconut. Rather prop it in the container so the fat end protrudes about two to four inches higher above the soil than the other end. Keep the starting medium moist, very moist, in fact. When sprouting starts at the end (and you will see it) it is time to transplant to a big tub with rich soil, with the coconut again protruding from the soil. Be sure the tub has excellent drainage facilities. The coconut likes lots of water but not water that just sits there. It must drain freely. (See the drawing on page 12.)

It will take about a year before the actual fronds open and the plant looks like a palm. When the fronds develop, it is time to cover the decayed portion of the nut. Use a good rich potting soil. In time the palm will decorate your room with lovely green, but don't ever expect anything more, like fruit, for example. That is beyond the power of the amateur gardener, or for a matter of fact, the advanced one.

Date Palm

Dates are infinitely good for you and loaded with all kinds of vitamins, and date palms are lovely indoor trees, so there are two reasons not to dismiss this venture with a shrug of the shoulders. Don't make the mistake of using pasteurized dates that have been steamed and preserved with chemicals. This is a mistake nutritionally speaking, too. Get the pure dates, that haven't been tampered with, at health food stores. After you have finished your date dessert, wash the pit or pits (it's best to start several) and put them in a starter mix. The time for germination varies, but it could be as long as two months, so don't give up in disgust. Keep the container in a warm place with good humidity. (Use the Baggie-on-a-stick method.)

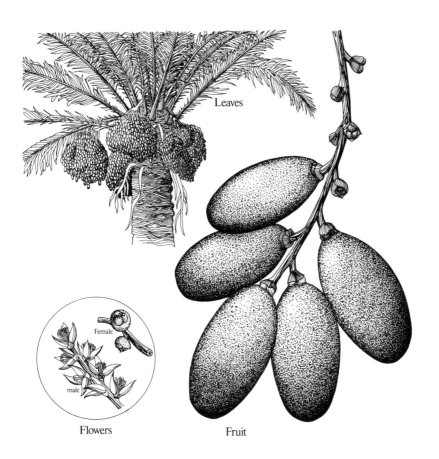

Leaves

Female

male

Flowers

Fruit

DATE PALM

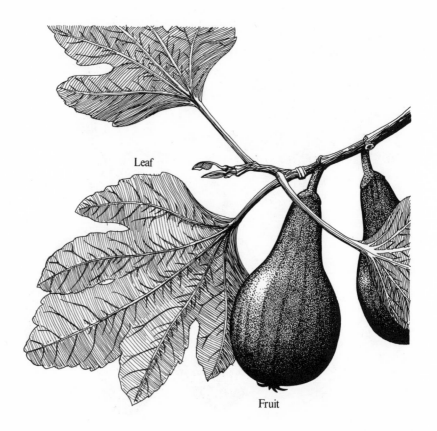

Leaf

Fruit

FIG

The seed cotyledon that acts as a reservoir for food for most plants usually comes from the top or above ground—not so with the date. It comes from the bottom, and it travels through the soil (like a root), coming up many inches later. When the root is about an inch long above the soil, it is time to transplant to a large tub of rich soil so the plant can grow on (hopefully). In a few weeks the root, or what was a root, or maybe *is* a root, will be joined by another, and presto, fronds! Your date is on the way. Give the plant plenty of sun, good moisture, and occasional feeding to keep it gracing your home. Unlike many plants mentioned in this book that remain medium and small, with the proper care the date palm can grow into a large plant—all from a little pit. Such miracles keep gardeners like me alive. The date is botanically called *Phoenix dactylifera*.

Fig

This tree grows in my backyard, and I have more figs than I know what to do with. Because the fig was so pretty outside, I decided I had to have one indoors, and since I couldn't move the fifteen-foot tree into my kitchen I had to come up with something. You can start seed from dried figs, but this is very difficult; it didn't work for me, so I cheated and took a cutting —and you can do the same. Get a cutting from a friend in California or wherever and start it in vermiculite as previously outlined in Chapter 3. Or if you want to invest money, small fig plants called evergreen figs are now available by mail

71

order. This is one time you can't have your cake and eat it, too. Once you eat the fig you eat the seed, too, so you have to get cuttings to start. No, this isn't a pit or a seed but a cutting, and I have taken a liberty. Please forgive it, but the fig is such a handsome plant I didn't want you to miss it.

The fig is a deciduous shrub or small tree and belongs to the mulberry family. It has lovely lobed green leaves and makes a splendid patio or indoor plant that actually grows with little attention, save for constantly moist soil and bright light. Its proper name is *Ficus carica*.

Guava

You can't buy a guava plant at nurseries; you must grow it from the pit. Called the sand plum by the Aztecs, the guava is delicious eaten fresh or as a dessert with cream. The hard pits must be clean of the fleshy fruit, so after eating wash the pit in warm water and get it started right away. Cover the pit with about one-quarter inch of soil and keep it warm in a bright place. The first shoots should appear in two to four weeks; if not, you've done something wrong or have a stubborn pit. The tree grows in an upright manner and can be pruned and trimmed without harm. Easily managed in a large tub (once the seedling has matured), the guava can become a decorative outdoor patio plant where temperatures don't fall below 55° F. Once it is actively growing, remove the suckers that form at the base and let three or four good branches grow; remove the rest. The guava is known botanically by the name *Psidium guajava*.

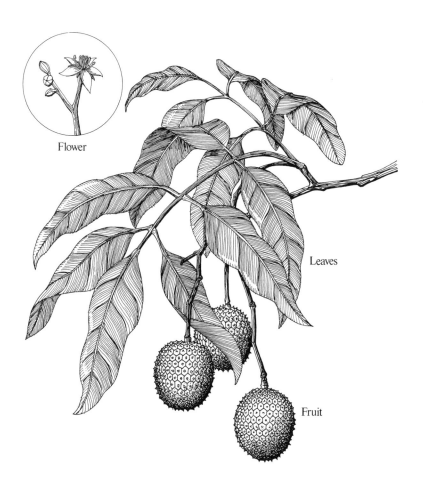

Flower

Leaves

Fruit

LITCHI

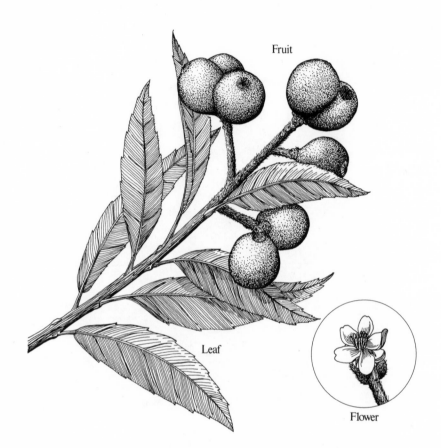

Fruit

Leaf

Flower

LOQUAT

Litchi Nut

You've eaten these, no doubt, in Chinese restaurants, where they're served either within a dish (succulent and sweet) or dried (sweet and dry). From southern China, the litchi has been around for more than 2,000 years. For your purposes, you'll need fresh litchis; the dried ones won't grow. So go to Chinatown to get them. The smooth-skinned pit, dark brown-black in color, is pretty in itself without planting, but once planted it produces a lovely narrow-leaved plant. Litchi is difficult to get started, so sow several seeds, one to a pot, in an acid-rich soil. Plant them about one-half inch deep. Keep them shaded and water them well. If you don't see signs of growth in two to three weeks, you've erred—it's probably a dried litchi fruit pit, which will do nothing. Keep litchis well watered and gradually expose them to bright light but never sun. This sweet delicacy is known in the plant world as *Litchi chinensis*.

Loquat

The loquat, another Chinese plant, should really be used more because the fruit is especially good just eaten out of hand or in chicken casseroles. The plant, too, deserves more attention because it's a lovely bold-leaved green beauty. The fruit resembles an apricot when it is ready for picking because of its orangy color. Wash and dry the seeds and then plant them one-half inch deep in good fertile soil that has ample drainage. With proper care and frequent repottings, the loquat can grow into an attractive tree. It can become part of the living-room decoration or be put in the garden where temperatures don't go below 15° F. Sometimes known as the Japanese medlar, the plant is formally known as *Eriobotrya japonica*.

Mango

There are several different mangoes and several ways of starting this plant. The lovely mango was cultivated in India 4,000 years ago (that should impress guests), and the plant itself is impressive, with leafy green foliage. The fruit, delicious for breakfast or dessert, is yellow-and-red with black specks, and generally kidney-shaped. A pit can be started (after you eat the fruit, of course) either by drying it first for a few days, or by soaking it in water for a few days, or by nicking its edge with a knife. In any case, first clean the pit by rubbing it with a stiff brush. (Wear gloves if you have sensitive skin; the mango does not like to be fondled; it can cause a rash.) Set the pit on end with the "eye" up and suspend it (as you would an avocado) with toothpicks in water or in starter mix so that the bottom inch or two is in the medium. Put the container in a warm, bright place; sprouts should appear in a month. Then transfer to rich soil in an eight-inch pot. Firm the soil around the collar of the plant but do not bury the stem. Keep soil evenly moist. The botanical name for the mango is *Mangifera indica*.

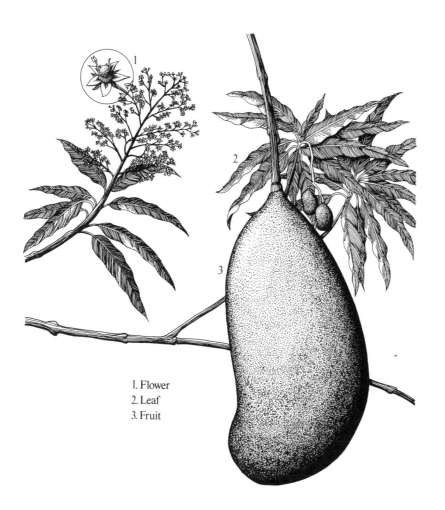

1. Flower
2. Leaf
3. Fruit

MANGO

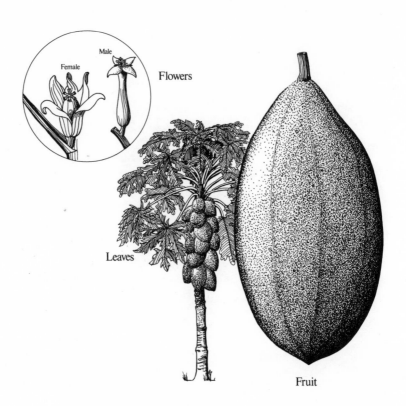

Female   Male

Flowers

Leaves

Fruit

PAPAYA

Papaya

Also known as pawpaw, this delicious fruit is good for you as
a dessert or as a digestive aid. Papaya flowers can be male, female, or hermaphrodite. In starting a plant, first wash the
seeds well and try to remove the sleazy outer coating; then
sow immediately. Cover with a scant one-eighth inch of soil
and keep in a warm, bright place. If germination hasn't
started in four to eight weeks, you've handled the sexes
incorrectly, in which case try a more delicate method of starting the perennial—sow in vermiculite and keep a plastic Baggie over the whole thing. Papayas are prone to damping-off
(see Chapter 6). Water copiously in the summer but not so
much in the winter. Repot the plants frequently, the first time
six months after you start, and then again six months later.
The papaya's leaves are scalloped and seven-lobed and remind
me of the popular houseplant called the umbrella tree. The
papaya can become, with care, an attractive houseplant, too,
and if you want to impress your guests, call it by its real name,
which is *Carica papaya*.

Passion Fruit

I had this pretty vine growing rampant outdoors. When it set
seed I immediately went to work growing it indoors. There

are more than 400 edible species of *Passiflora*, but only some provide the fruit used for jellies or desserts. My passion fruit was a nonedible kind, but it still produced a lavish plant. Leaves are large and deeply scalloped. Passion fruit can be started from seed as soon as it's available in the yard. Plant seed in a light soil and provide high humidity and warmth (78° F.). Germination should start in about a month or less. When the seedling is two to three inches tall put it in an individual pot. Keep it in a bright place with moderate temperatures; avoid extremes. A lovely, leafy plant.

Persimmon

If you've ever seen the lovely persimmon tree, you'll want a pot plant for the kitchen windowsill. It has ovate green leaves, and its bright orange fruit is eaten fresh or cooked. Germination is erratic and may take from two to ten weeks, depending on how you handle the seed. For best results and quick pit n' pot gardening, put the seed in some sphagnum moss after you've eaten the fruit and store the seed and moss in a closed bag in the refrigerator for ninety days. Then sow the seed and cover with a shallow layer of soil; give it warmth and bright light and keep your fingers crossed. If all this sounds like too much of a struggle, remember that few people can boast a persimmon tree on their windowsill. The persimmon is called *Diospyros kaki.*

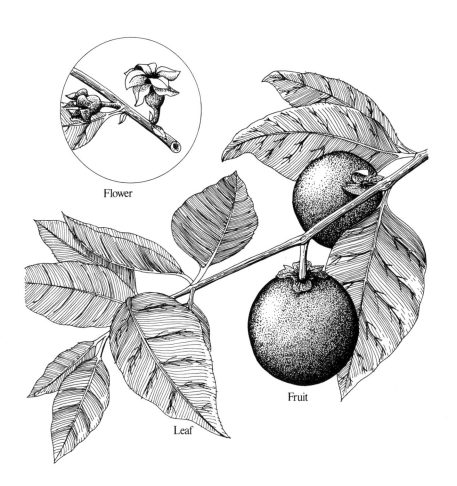

Flower

Leaf

Fruit

PERSIMMON

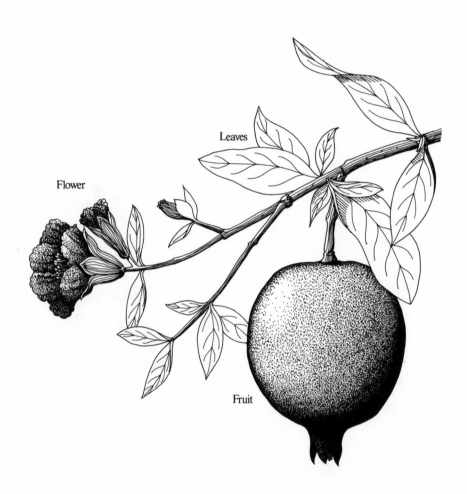

Flower

Leaves

Fruit

POMEGRANATE

Pineapple

Everyone knows that the pineapple has no pit, so what's it doing in this book? This is author's liberty—I happen to like the leafy pineapple plant. You can't find a better or more undemanding houseplant than this member of the bromeliad family. It can be started easily by slashing off the crown of the fruit (the part with the leaves) about one inch below the base of the leaves. Let it dry to prevent rotting of the stem after planting. Put the crown in a pot of sandy loam and place container and all on top of the refrigerator. The pineapple needs good bottom heat to get it going. Give the pineapple good drainage and frequent light waterings rather than occasional thorough waterings. If you're not frugal like me and want to buy a plant, they're available in florist shops. Don't get the miniature one, though; it's temperamental. Ask for the real one by name: *Ananas comosus.*

Pomegranate

Many people are probably familiar with the small pomegranate, *Punica granatum nana,* sold as a houseplant. But you can also grow the standard pomegranate, *P. granatum,* from seed. Introduced into England from Spain in the sixteenth century, this is a delicious fruit and makes a handsome indoor plant.

The fleshy substance around the pomegranate seed is what is eaten; it is cooling, sweet to the taste. Once you have eaten the goodies, take the seed and allow it to dry a few days. Then germinate the seed in shallow pans of vermiculite covered with a plastic Baggie to assure good humidity. Mold may accumulate on the seed, but don't panic—this is a symbiotic relationship necessary to germination. In about two months (if you're lucky) the seeds will crack open and growth starts. Now transfer the seedlings to individual pots of rich soil—equal parts potting soil and humus. Cover the seed and the taproot with the soil and in a few weeks leaves should appear. Give the plant warmth and sun; the soil should be evenly moist, never soggy, and the pomegranate dislikes high humidity.

You can also separate seed from pulp (the part we eat) by running the fruit through a sieve and rinsing away the fleshy part. That is, if you don't like to eat pomegranates.

Although eating and starting the seed of this plant is very possible even for the novice, getting the plant to bloom is another story and would require several more pages. So be content with the lovely foliage; it is a delightful pot plant.

Rambutan

This fruit, native to Malaysia, is beginning to appear in specialty markets. It belongs to the same family as the litchi and is grown in the same way. The fruit eaten—and it is good—is the white fleshy aril surrounding the single seed. It is sweet and acid and can be eaten raw, or it may be stewed. The seed

Leaves

Flower

Fruit

RAMBUTAN

has to be started in high humidity and good warmth in a sandy soil mix kept evenly moist. When germination occurs (and this may take many weeks) and green growth shows, transfer the plant to a tub with rich soil. Keep evenly moist and warm. The rambutan is a leafy branching plant that makes a fine, quite distinctive indoor accent. Although you might have to search for the fruit, it is worth the time. Botanically, this plant is called *Nephelium lappaceum*.

8.

Houseplants

HOUSEPLANTS have always been my favorites. When I first started adding some greenery to a drab Chicago apartment many years ago, I rarely bought plants; I'd start slips or cuttings from friends' plants, or when I could get them, I would pop seeds in a pot and hope for the best. I had many failures and only some successes. Now, some twenty years later, I find that growing my own window plants from seed still makes good sense. In terms of cost and care this beats buying plants, and many of the more unusual and beautiful plants that I want are simply not available easily (begonia species are a fine example). But somehow I always manage to secure seed from a friend's greenhouse plant or from some nursery source. You, too, can do the same thing; ask and inquire.

If this frugal way of growing houseplants seems like a

bother, think about the immense enjoyment in growing something from a tiny pit or seed and nurturing it to perfection. A plant you grow yourself is like a child—you won't forget to take care of it. The sense of satisfaction and solace to the soul is worth the extra effort and time required to find seed. So if people ask you why you grow houseplants from seeds when they're readily available at local nurseries, set them straight. Tell them that your taste in plants is for the unusual and you seek something besides the ubiquitous philodendron; a coffee plant or a lovely fibrous begonia is unique and generally not available from most suppliers.

When you start growing your own houseplants don't think you'll be able to put a nursery shingle outside your house in a few months or years. You won't. But you will be able to grow the plants you really want for your own personal enjoyment. Quite frankly, some houseplants are difficult to start from seed; others are easy, but experimentation is half the fun, so whether you get houseplant seed free or whether you buy it in packets (very inexpensively), there are rewards beyond financial compensation.

Care of Houseplants

Assuming that your seedlings are going to grow into flourishing plants (and most will whether you read this book or not), you'll then want to know how to take care of your mature plants so that they'll become long-lived additions to your home. Generally, most houseplants will get along just fine in

average home temperatures of 70° to 78° F. during the day
and ten to fifteen degrees cooler at night. Indeed, the plants
need the drop in temperature to assimilate foods. Humidity
may present more of a problem. The amount of moisture in
the air affects people and plants. If the air is too dry, people
suffer from sinus problems and plants suffer, too; they trans-
pire more quickly than they should through their leaves, and
in a few months the plant is lost. Fortunately, many houses
have humidifier systems as part of the heating operation. Home
humidity should be between 30 and 40 percent for plants (and
people). A small hygrometer in an inconspicuous place in the
living room will measure humidity. If you don't have a mod-
ern humidifier system and humidity is low, you'll have to do
some work to keep your plants healthy. This work involves
spraying plants with water a few times a week, and to ensure
that they get enough humidity you might want to set them on
gravel-filled trays. Plastic or metal trays are sold at nurseries,
or use clay saucers and fill them with gravel. Set plants on top
of, not in, the gravel, and keep an inch of water in the gravel
container.

Some people like fresh air, and other people can't tolerate it, but plants do prefer a good circulation of air because in
stagnant situations they may perish. A good circulation of air
means just that—no drafts. Drafts are death to most plants
and can ruin them overnight, so keep plants away from hot-
air registers or air-conditioning ducts. To provide the gentle
flow of air plants like—even in cold weather—open a window
a bit somewhere in the room but not directly near the plants.

A good potting soil for plants has already been described,
but any potting soil after a time will eventually be leached of
all its nutrients, so it will become necessary to feed plants in

confined containers. All you need is one bottle of plant food; select one that has the numbers 10-10-5 on the bottle. This means it contains nitrogen, phosphorous, and potash, in those percentages. These elements are necessary for good plant growth. You may also want to feed your plants a few times a year with fish emulsion, smelly but excellent stuff.

New plants and ailing ones don't need additional food; plant roots can't accept it. It's a waste of money and might ruin the plant. Start feeding mature plants in the spring; do it with every other watering, and continue through the summer. Taper off in the autumn to a once-a-week application, and in the winter don't feed them at all. Once a month—and this is important—take plants to the sink or shower and soak them thoroughly to leach out accumulated salts that can harm plants.

How much water to give a plant varies with the plant being grown, the kind of pot it is in, and where you live. Because of these many variables it would be impossible to put down rigid watering schedules. In general, never allow any plant to become completely dry or completely soggy. If soil is too dry, plant roots become dehydrated and growth stops. Continuously wet soil becomes sour and causes roots to rot. Try to maintain an evenly moist soil for your plants. If this entails too much work, water plants heavily, but don't water them again until they're almost dry to the touch. When you water, do it thoroughly. It's harmful if only the top few inches of the soil become moist and the bottom of the soil, where roots rest, is bone-dry. Do use room temperature water rather than icy cold water that can shock plants.

In a year or two after a plant graduates from seedling to living room, bedroom, or wherever, it has to be replanted because the soil will be depleted of nutrients and the plant will thus need a general pickup. Select a clean pot (or thoroughly wash an old one) and fit an arching piece of crock or potshard over the drainage hole. (*Always* use a pot with a drainage hole.) Over the crocking spread some porous stones and some charcoal bits (now sold in packages) to keep the soil sweet. Put in a mound of soil, center the plant, and fill in and around it with soil. To settle soil and eliminate air pockets, strike the base of the pot on a table a few times, and then with your thumbs firm the soil around the plant collar. Don't bury the collar of the plant too deeply or rot will occur. Leave about an inch of space at the top between the pot rim and the soil so you can water plants easily. Soak plants thoroughly after re-potting and keep them in subdued light for a few days.

Light, mentioned here last, should probably be discussed first because without adequate light few plants will thrive. Generally, most indoor plants will get along just fine with a few hours of daily sunlight. Few will prosper in direct sun for many hours, so put plants in a bright place—an east or west window. A southern exposure can become brutally hot in summer for most indoor plants. Once a month turn plants a half turn so they can benefit from light from all angles.

Abutilon

The flowering maple, abutilon, not to be confused with the tree, is a member of the mallow family. The plant has attractive lobed leaves and bears abundant orange or yellow holly-

91

hock-shaped flowers. This houseplant is rarely sold, so seed sowing is your only way of having it on your windowsill.

Put seeds in shallow pots containing a fertile soil-and-sand mix and give them warmth, at least 75° F. A good place for the container is on top of the refrigerator. Germination is fast, usually within three weeks, and the little plants will grow rapidly. Prune and trim them when they're a few inches high so they become bushy rather than leggy, which is their tendency. Then grow them in sun and warmth, and never forget to water them because they wilt quickly without ample moisture.

Agapanthus

Here is a plant rarely seen in most parts of the country but certainly desirable for the small window garden. With grassy leaves, agapanthus bears crowns of tiny blue flowers. This member of the lily family can be started from seed in early spring. Use a rich soil and sow seeds lightly on top of the soil. A regular home temperature of 65° F. is fine. Germination starts in twenty to thirty days. Keep transplanting seedlings as they mature; eventually the plant will need a large tub.

In the spring and summer keep the plants wet, but in the winter they appreciate only enough moisture to keep the leaves from drying. Agapanthus can grow into a fine indoor plant that will bear flowers every year. You can also divide large plants to get new ones; indeed, you might end up after a few years with more agapanthus than you really want!

AGAPANTHUS

BEGONIA

Begonias

These popular houseplants include a host of different kinds for indoor growing. Tuberous-rooted ones grown from tubers are great favorites, but some of the fibrous types are easily grown from seed and include many lovely decorative plants. Start begonia seed in pans of vermiculite rather than in pots. The seeds are minute and shouldn't be covered with growing medium; merely toss them on top of the vermiculite, give them warmth (75° F.), and enclose them in a plastic tent or some other closed container so that they have good humidity to ensure germination.

You should see the first signs of life in about a month, although germination varies with the specific begonia grown. When seedlings are up, remove the plastic enclosures and water them with a fine mist until they're really growing; then regular watering is satisfactory. Transplant seedlings into individual pots and give them only bright (not sunny) light all year except in the winter, when most begonias can tolerate full sun. Keep plants evenly moist, never soggy.

Cacti

I strongly recommend you start cacti from seed because good cactus plants (or even bad ones) are difficult to find unless you want to settle for the dime-store varieties that can run into money if you buy more than two. Some cactus seed is fine; some is large. Large seeds should be planted their depth in the growing medium; plant small ones just lightly covered. Use a starting medium of equal parts vermiculite, perlite, sand, and sphagnum. Germination is erratic; some plants may start into growth in two to three weeks, but others will take a few weeks longer. Keep your eyes open and transplant seedlings when they are big enough to handle.

Once you have mature plants, don't make the common error of not watering cacti in the fall and winter. It's true most rest in nature, but in your home they'll need some water (about twice a month) during the dull days. Keep them happy.

Here are a few cacti to try: *Ariocarpus, Lobivia, Parodia,* and *Rebutia.*

Coffee Plant

This plant is occasionally available at shops, but more frequently you'll have to pit n' pot to get it. The problem is where to get the pit, or bean. Dried ones won't work; you've got to have fresh ones. Small specialty shops in large cities now roast their own coffee, and these are the places to get a fresh bean or two. If at first you don't succeed, try again, because the rewards are worth it. The coffee plant has lush green leaves and in blossom is a beautiful sight. The fragrant white flowers are in clusters followed by clumps of green cherry-ripe fruit turning golden brown and finally red.

ARIOCARPUS

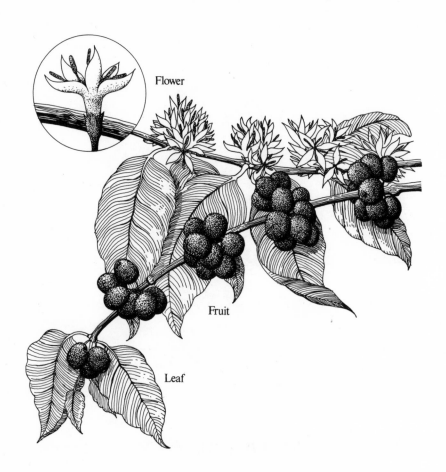

Flower

Fruit

Leaf

COFFEE

To start your coffee plantation, use a sandy soil and put beans at least an inch below the soil line. Keep the temperature high, at least 85° F., to ensure germination. The plant grows rapidly once started and will need some sun and frequent misting to become a healthy adult. Warm, moist conditions are the key to success with the mature plant, but not excessive heat. The ideal temperature is 70° F.

Crossandra

Here is a quick-growing plant from India. The orange flowers are exceedingly handsome, and even small plants bear them throughout the year. The leaves are glossy dark green. To get crossandra started, use a good rich potting soil, sow seed, and keep warm (80° F.). Plants should start to sprout in about a month. Crossandra is a delightful window plant and once established will be with you for years. Give mature plants winter sun, bright light the rest of the year, and water them copiously because they like to be moist.

Geranium

The geranium is a good example of a plant that blooms about six months after starting. Seeds germinate best at 65° F., requiring about two to three weeks' time. Keep them evenly moist and keep shifting seedlings from one pot size to the next larger one as they mature. This is little to ask in the way of work for all the plants you'll get. There are many different kinds of geraniums and many different flowering habits, but no matter which ones you get from seed, in the winter give mature plants ample sun or they'll forever remain uncooperative. There's nothing sadder than a bunch of geranium leaves without their pretty flowers.

Gesneriads

This is a huge family with some stunning plants that are not too well known (except for African violets) but which deserve more attention. But achimenes, episcias, gloxinias, kohlerias, and smithianthas, for example, are all fine houseplants. Most plants in this family have rather soft, velvet-textured leaves and bear lovely, brightly colored, handsome flowers, some in summer, others in fall—a whole bouquet of blooms.

Plant seeds in vermiculite or in a soil-and-sand mix in trays or pots. Transplant seedlings when true leaves are up and then in a few months repot them again in fresh soil. Specific directions are given for each plant below.

ACHIMENES. These are fine summer-blooming plants with flat-faced flowers. The seed is fine, so don't cover it when sowing in containers. But do give seeds lots of heat (85° F.) to get them started. Mature plants need plenty of sun and water.

EPISCIAS. These are mainly trailing plants, with multicolored leaves. The foliage is beautiful by itself, but the

GERANIUM

EPISCIA

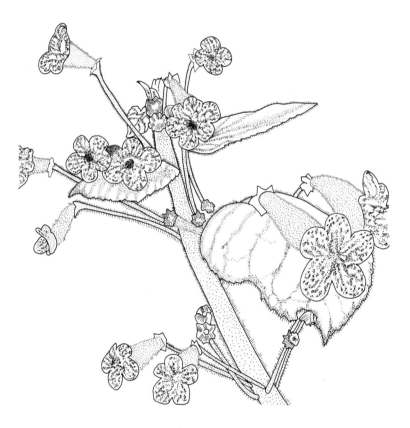

KOHLERIA

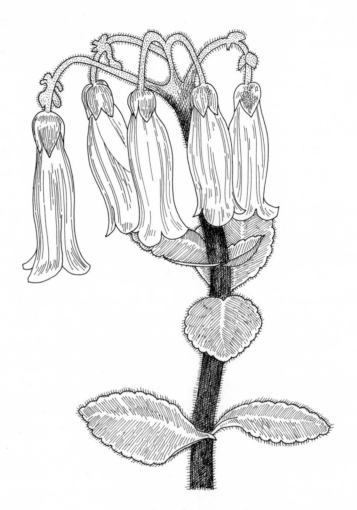

SMITHIANTHA

plants also bear bright-colored flowers. Start the seeds in a warm place (about 75° F.) and press them very lightly into the starting mix. Germination starts in about a month, and once up, the seedlings can be planted in individual pots. Mature plants never want to be soggy; keep the soil evenly moist. A slight drop in temperature at night will keep episcias thriving.

GLOXINIAS. If you're too frugal to pay the price for these gorgeous gift plants at seasonal times, start seed; you'll be able to supply the neighborhood. Press the seed very lightly into the growing medium and keep the temperature at 70° F. Tiny plants should start in about three weeks. Then put them into individual containers and give them good light, but keep mature plants cool (65° F.) and evenly moist, never soggy. This plant produces tubers. After the plant blooms (in about six months) let leaves die back naturally. Store for two months in a paper bag in a cool place and then start tubers all over again.

KOHLERIAS. Handsome plants with bell-shaped flowers that make a fine color display. Start seeds in shallow pans of vermiculite at about 75° F. Keep evenly moist in a somewhat shady place. Start feeding when the first leaves show; transplant when the second set of leaves appears.

SMITHIANTHAS. Commonly called temple bells, the smithiantha has dark, velvety leaves and bears an orange-and-yellow flower spike. Seed should be started at 70° F. in vermiculite in shallow pans placed in bright light. Keep moist. When plants show leaves, start a mild feeding program, say every other watering. After blooming, allow to rest for about a month with less water.

105

KALANCHOE
"Jingle Bells"

Kalanchoe

This fine plant with fleshy leaves and bright red or orange flowers has become very popular as a Christmas gift. Technically called *Kalanchoe blossfeldiana,* this plant and its varieties (there are many) is easily started from seed in the spring. Put seed in shallow pans of vermiculite at about 75° to 78° F. for germination. Be alert for damping-off disease, which can occur if the humidity is too high. Keep evenly moist and transplant when leaves are two to three inches high. Put seedlings in a soil mixture of equal parts sand and rich humus. Keep the soil evenly moist and place the plants in a bright window.

Saintpaulia (African violet)

African violets are gesneriads but so popular that they deserve a section of their own. Everyone loves these denizens of the jungle. Although they start easily from seed, once they're mature, the plants take time and patience to bring to perfection. To start them, put the dustlike seed in a starting medium in warmth (70° F.) in a shallow container and cover with a sheet of plastic or glass. Scatter seeds lightly over the soil in the dish or pot. Keep in warmth (75° to 80° F.) in a bright light. Look for sprouts in about two to four weeks, although the time varies somewhat depending upon the parent varieties. Keep the

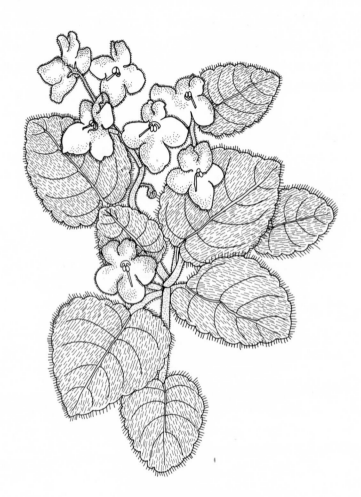

AFRICAN VIOLET

growing medium just damp, never wet. Remove the plastic or glass cover if you notice moisture condensing heavily. Leave the cover off a few hours a day. Watch new plants carefully for damping-off; if you see gray mold on the soil surface, apply fungicide (Dexon is effective). When seedlings crowd each other, transplant them to individual pots (they should be at least two inches tall).

9.

Wild Plants

OUR NATIVE PLANTS offer a wealth of beauty. At nurseries we generally buy hybrid varieties, but seed from them won't produce true to its parent. However, our wildlings, untouched by the hybridizer, will produce identical plants and with far less trouble than you think. Watch for the time when seed is mature and ready for planting. There's no harm in taking some seed to sow in a pot for your indoor garden but *never* take plants. Such plants as elderberry and bloodroot and dozens of others are hardly known by most people, so here's a chance to really get acquainted with your native flora, with no cash outlay on your part.

However, don't be misled into thinking you can move nature in one fell swoop. Quite frankly, wild plants provide a pail of sand to a high shelf, or lifting a pail of water from the floor to the sink.

Purpose: To discover that cars go faster and more smoothly over a surface where there is little friction.

Problem: Do cars travel better on rough or on smooth roads?

Materials: Toy cars; a smooth waxed board on which to roll the cars; a board which has been roughened by pouring onto it a mixture of plaster of Paris and pebbles.

Activity: The children will like to engage in a play which involves allowing the toy cars to run down the smooth board and the rough board. They will determine that the cars go faster on the smooth surface than they do on the rough surface. Usually, the boards will be raised and lowered to discover differences in car speed also.

Purpose: To become aware that stairs help people walk inclines more easily.

Problem: Why are steps built (instead of ramps) to go up or down to the next floor?

Materials: Narrow strips of construction paper.

Activity: The children will enjoy learning to fold the paper accordian-style to represent a flight of stairs and then have their fingers do the walking. Later, to complete the experience, they may walk real stairs.

Purpose: To learn how gears move.

Problem: How do gears and wheels work for us?

Materials: A collection of gears and wheels; several old toy music boxes with their covers removed so that the gears may be observed in action.

Activity: Encourage groups of children to examine and play with the assorted gears and wheels. Some boys and girls will notice the difference between gears and wheels. Wind up the music boxes and let the children discover how the gears and notched music drum help produce the sound.

PLANT LIFE

Purpose: To notice the differences in weeds.

Problem: Do all plants look alike?

spot of color—that's about it. Unless you have a garden and after a certain time can start a few plants outdoors, enjoy the seedlings indoors. The best time to get the seed is about a month after the flowering period. Seed capsules and pods can be taken from the stems without hurting the plant. Then crush out the seed by hand or tap the pods on a newspaper. Don't avoid seeds with fleshy coverings; remove the coverings by soaking the seed in warm water for fifteen minutes. Then put the seeds in a mesh smaller than the seed and scrub off coatings with a stiff-bristled brush. Store seeds in glass bottles if you can't use them immediately. Most wild seed ripens and falls to the ground four or five weeks after blooming time.

Since this book is for the home gardener, it doesn't seem necessary to define details about how seeds of wild plants differ in specific requirements for germination. But some things mentioned about them in Chapter 3 deserve to be said again here because they are important in getting seeds to germinate —some seeds are best sown as soon as ripe, others need a ripening or cooling period for germination to occur. The latter kind require temperatures of 40° to 45° F. for 60 or 90 or 120 days. So, once sown in the starting mix, put seed containers in a plastic bag (Baggies are fine) and store in the refrigerator. When the incubation or sleeping time is over (these times are mentioned with the plant descriptions that follow) move seeds to a bright, warm place. Keep the plastic covering on the container until you see tiny leaves appear. Remove the plastic only if too much condensation occurs (and you will see this on the inside of the plastic). Sowing the seed of wild plants is the same as for starting any seed, as explained in Chapter 3, and the types of starting mixes are the same, too.

Bloodroot

This is a lovely, lush green plant with dark-green lobed leaves and beautiful white flowers. I grow it for its beauty as well as for its fine red dye for yarns. Bloodroot is generally unavailable from suppliers, so you must grow your own. You'll find the plant in shady woodsy areas, occasionally along roadsides, but don't pick; it's scarce. Wait for seed and then gather it as previously explained. Sow the seed in a starter mix in shallow trays or in pots in late fall. Cover the container with plastic and put in the refrigerator for a long sleep—about 90 to 100 days. In early spring, move the containers to a bright, warm place. Remove the plastic covering as soon as sprouts show. When leaves are up, repot in soil and grow on in a bright window. Repotting may be necessary again in a few months— a larger pot and new soil. If you are going to try to grow the plant indoors, find a shady, moist spot for it; outdoors, it can go under trees or in any protected area. Bloodroot is properly known as *Sanguinaria canadensis.*

Butterflyweed

This fine shrub, a member of the milkweed family, produces lovely clusters of orange flowers. Rarely available in nurseries, butterflyweed can be nurtured indoors until it's mature; at that time move it to the garden, although it is so pretty you may be reluctant to part with it. Gather seed (thin, brown,

BUTTERFLYWEED

DOGWOOD

and pear-shaped) when the big pods are turning brown. Before saving, or sowing, remove the silky parachutes. Sow seed in trays or pots of vermiculite in February and keep shaded; provide good humidity—use plastic over the container. When germination starts and you see leaves, remove the covering of plastic. Plants generally sprout in about four weeks. Then repot in a container of sandy soil and put the plant in a bright window. The butterflyweed (which does attract butterflies outdoors) is known botanically as *Asclepias tuberosa*.

Dogtooth Violet

This plant is found in dry or moist open woodlands. It has miniature lily-shaped flowers and speckled leaves. Seeds are pale brown and pear-shaped and germinate readily. Sow them in trays or pots as soon as you pick them; use a starter mix and give seeds coolness (55° F.) at first and then gradually increase heat by moving them closer to windows. Repot in individual containers of soil when plants are two to four inches tall. Plant outdoors in the garden or try the violets at a cool, shady window. Technically, dogtooth violet is known as *Erythronium giganteum*.

Dogwood

It's not too difficult to have your own dogwood on the windowsill where it will grow for some time but not indefinitely.

115

Eventually, it needs outdoor conditions. This lovely ornamental tree has rosette flowers in spring and deep scarlet leaves in fall; the berrylike fruits are red. Seed requires a long cooling period of about 160 days, so sow seed in starter mix (sphagnum seems good here), cover with plastic, and put in the refrigerator. In spring remove the container and place in a bright, warm place. Leave the plastic on for a while to see if germination starts, or remove the covering (as for many plants in this section) if too much condensation occurs. This could cause damping-off. Now if seedlings don't come up, don't panic. This is a tough one and sometimes nothing happens until the second spring. Throw them out? No! Just set them aside on a back porch or in a garage or somewhere convenient and keep the mix slightly moist. Come spring again, give them warmth and more moisture. Once leaves show, repot in fresh soil and then in a few months put in a larger pot with fresh soil. Sound like work? It is, but not everyone can have a dogwood tree in his house. Dogwood's botanical name is *Cornus florida*.

Elderberry

This deciduous shrub, with its large and lacy white flowers, makes a pretty houseplant for a time, but don't expect it to last forever unless you can put it in the garden. The purple-black berries are a delicacy for birds, so you'll have to gather seeds quickly and early when they're ripe. Separate the fruit from the pits (as described on page 111) and sow the seed immediately in fertile soil in a shallow clay pot. Cover the pot

ELDERBERRY

with a Baggie; keep the temperature warm and the humidity at a good level. If you're successful, this plant can go into your garden when it gets to be a good size. As a treat, make elderberry wine, or dip the blossom heads in batter and fry in deep fat. Delicious! *Sambucus canadensis* is the proper name of this denizen of moist meadows.

Hackberry

This is a rather common tree in north temperate zones; if you run across berries on your outdoor hikes, gather a few for planting. The berry is at first yellow or red but turns purple-black when ripe. Start the hackberry seeds in trays or pots of starter mix in spring. Cover with a Baggie and put into the refrigerator for the winter rest of sixty to ninety days. In spring remove the seeds from their deep sleep and place at windows where it is warm. When plants are a few inches high, transplant them into pots of rich soil for growing at the window, or put them directly into the ground in the garden. The leafy plant is handsome once it starts to grow and makes a nice indoor plant for a while but not forever. It belongs outdoors. The hackberry is officially known as *Celtis occidentalis*.

Marsh Marigold

This is one of our loveliest wild flowers and so easy and interesting to start from seed. The plant has healthy dark green leaves and bright yellow buttercup flowers. Marsh marigolds are found in open woods or swamps. The brown seeds are elliptical and pear-shaped. Pot immediately one-quarter inch deep in a rich mulchy soil in a glass terrarium. Then sit back to enjoy this real beauty. (Plants can also be started in the garden.) Marsh marigold is known botanically as *Caltha palustris*.

Natal Plum

This is a lovely shrub with small, dark-green leaves and edible berries. Because it has recently become popular as a houseplant I didn't know which section to put it in but because I have known it as a wildling for years, decided it should be here. To have your own Natal plum, wash pulp from the fruit and sow the seed immediately in trays or pots of vermiculite. Give the container warmth (about 78° F.) and bright light. You can put on a plastic cover, but generally it isn't necessary, and the Natal plum should start growing in about ten to fourteen days. This is really an easy one to get started so do give it a try; it's an inexpensive way to have a fine indoor plant. Keep repotting the plant in fresh soil as it grows on. The official name for the plant is *Carissa grandiflora*.

Prickly Pear

The numerous species of these cacti are native to the Southwest. The plants are tree-shaped, with beaver's-tail leaves covered with spines. I include the prickly pear here because it really is an easy-to-grow plant and can in time become a fine indoor accent plant. Flowers are large and showy, and a healthy plant will bloom indoors. The seeds are flat and somewhat large. After you have collected seed, put it in starter mix in trays or pots and cover with a Baggie. Store in the refrigerator until spring. Then remove and place in a bright, warm window. Remove the plastic or keep it on to assure good humidity; however, when plants start to sprout, remove the plastic covering. When the seedlings are ready for transplanting, put them in sandy soil in a tub or pot. With prickly pear, germination is erratic and may take many months or just a few, so have patience with this one. Prickly pear is known as *Opuntia basilaris* to the botanists.

Serviceberry

This attractive shrub or small tree of the genus *Amelanchier* has toothed oval leaves and applelike small fruits that are purple-red to black when ripe. Flowers are white and appear in April (but not indoors). Almost any species of this native shrub can be used. You'll find this wildling along stream banks and on moist hillsides.

For your indoor growing gather the seed when ripe and place in starter mix in pans or pots; put Baggies over the containers and place in the refrigerator for 60 to 100 days. In spring, remove the containers to a bright, warm window so the

PRICKLY PEAR

SERVICEBERRY

seed can germinate. When leaves are up, remove the plastic coverings; seedlings are ready for transplanting. Pot them in rich soil.

Snowberry

This creeping evergreen shrub is known as *Chiognemes hispidula*. It can easily be grown from seed. It's best to start snowberry in the autumn; you'll find the plant then in mossy woodlands or any mucky place. The small leaves are dark glossy green above and brown underneath. Sow seed in sand and peat or starter mix in a tray or pot and keep cool for about 120 days (45° F.). Put it in an unheated garage or in the refrigerator. In spring bring it into the kitchen, where germination should start. Keep the soil evenly moist. The berries are edible as they come from the plant or as a dessert, but indoors you'll have to be content with only the leafy beauty of the plant.

Spanish Bayonet

This makes an excellent houseplant and yet to my knowledge is rarely seen. Leaves are dark green, spear-shaped, and grow in a lovely rosette. The plant eventually grows quite tall and makes a fine window accent. The large black pear-shaped seeds are in protective capsules; crush the outside to get the seed. Start seed immediately in shallow pots of starter mix. Give warmth and bright light and cover the container with plastic to assure good humidity. Seed should germinate in thirty to forty-five days. When fledglings are a few inches high, put them in tubs of rich soil. The plant is officially known as *Yucca aloifolia*.

Sunflower

This annual or perennial plant, with its nodding head and gigantic flower, is well known. You'll find sunflowers on the plains, prairies, and in the valleys of the West. The seeds have varied food uses. To start your own windowsill sunflower, embed seed anytime, one-quarter inch deep, in a pot of rich soil or starter mix. Keep warm (70°–80° F.) and shaded. Plants should show growth in about two weeks. This is an interesting window plant simply for the sake of seeing how fast it can grow; eventually it will need space because it can reach fifteen feet. The species name is *Helianthus annuus*.

Trailing Arbutus

This evergreen creeper with fragrant pink flowers is generally found in moist and acid soils in forests or woods. The oblong

SNOWBERRY

SUNFLOWER

pitted seeds are tiny (pea-size) and are borne on a waxy receptacle and enclosed in a hairy capsule. When the capsule is ripe, it splits into five parts, exposing the seed. Sow seed in shallow pans or trays of sphagnum in late fall and keep covered with plastic in the refrigerator for sixty to ninety days. In spring move the containers to a bright window; keep the plastic on to assure humidity, or remove it if too much condensation occurs. When plants are at the seedling stage, move them into pots of soil. Keep trailing arbutus indoors as a pot plant for at least a year or more before planting outdoors (or keep it inside indefinitely if you like). This plant is excellent for terrariums, where it'll need rich soil and shade at all times. Trailing arbutus is known botanically as *Epigaea repens*.

Trillium

When I lived in Illinois, in early spring my property was a canopy of white trillium, gradually turning pink. This woodland inhabitant likes shade. The seeds are reddish brown and oval and can be started in trays or pots of starter mix in late winter. Cover with plastic and put in the refrigerator for sixty to ninety days. Then remove from the refrigerator in spring and give them gradual heat; take off the plastic covering if too much condensation occurs. When the seedlings are big enough to move (about four to five inches) either try them in pots of rich acid soil in a shady place in the kitchen, or if you have a garden, transplant them there in a rich mucky place where again there is ample shade. Trillium is a good native to grow because it is becoming scarce in nature, and it's rare to see it outdoors much anymore. There are many species of trillium, and the growing method above should work for most of them.

Wild Geranium

This hardy perennial, known for its lovely flowers, is found in wooded areas and in the shade of trees or shrubs. The seeds are dark and oval and appear on the lower end of stiff, strap-shaped fruit segments. When ripe, the straps curl and hurl the seeds into the air. Seeds germinate quickly and readily in any starter mix. When a few inches high, transplant into pots of soil. The wild geranium is an overlooked but excellent window plant for color accent. Water thoroughly and deeply all year, and give it sun.

Wintergreen

This low-growing plant has leaves arranged in whorls and produces red berries, which, when crushed in your hand, have a strong wintergreen scent. Seeds are pear-shaped and reddish brown. When you collect seed, place it in starter mix in trays or pots and cover with Baggies. Now put the containers in the refrigerator for several weeks, then remove them and keep at

WILD GERANIUM

WINTERGREEN

room temperature for several weeks. Repeat the process again, and then once more, or until spring is on the way. If all this sounds like too much trouble, it is, but it is not impossible. Remove the plastic covering when the weather is stable and let the seed germinate in warmth (78° F.) in a bright window. When fledglings are up, move them to pots of rich acid soil, or of course, you can plant them directly in the garden. Wintergreen is properly known as *Gaultheria procumbens*.

List of Suppliers

SEEDS OR PITS for the kitchen plants mentioned in this book are generally available in season (sometimes in all seasons) at the fruit department of grocery stores or specialty food shops.

If you do not want to or cannot gather your own wild plant or houseplant seed, you may order seed from the following sources:

Leslie's Wildflower Nursery
30 Summer St.
Methuen, Mass. 01844

12-page catalog—25 cents
(wild flower seed)

Lounsberry Gardens
P.O. Box 135
Oakford, Ill. 62673

Catalog—25 cents (wild
flower seed)

LIST OF SUPPLIERS

Clyde Robin   Catalog—1 dollar
P.O. Box 2091   (wild flower seed)
Castro Valley, Cal. 94546

George Park Seed Co.   Catalog—free (houseplants)
Greenwood, S.C. 29646

W. Atlee Burpee Co.   Catalog—free (houseplants)
Hunting Pk. at 18th
Philadelphia, Pa. 19132

Index

Numbers in italic refer to illustrations.

135

INDEX

139